Bibliographic information published by the German National Library:

The German National Library lists this publication in the National Bibliography; detailed bibliographic data are available on the Internet at http://dnb.dnb.de .

Imprint:

Print and binding: Books on Demand GmbH, Norderstedt Germany
ISBN: 9783668676176

This book at GRIN:

https://www.grin.com/document/418540

Hans Georg Schrey

Effect of Uneven Cooling on Performance of Air-Cooled Condenser

GRIN Verlag

Title **Effect of Uneven Cooling on Performance of Air-Cooled Condenser**

Author Hans Georg Schrey

Keywords: Air-cooled Condenser; ACC; Cold Spot; ACC Control; Dephlegmator Sizing

Contents

Abstract

Air-cooled condensers ("ACC") operating in vacuum are widely used at the cold end of contemporary thermal power plants. Proper functioning of the condenser is paramount for power plant efficiency.

To adapt to changing process conditions because of changed ambient air temperature or power station load ACC streets are either taken out of or into service or, module fan speed settings are selected appropriately. Even if all fans are running at same speed local variations of cooling air flow may arise as a consequence of e.g. fan location or local wind impact. The designer must ensure that no negative effect (i.e. no cold spot) may evolve at the steam side of the ACC caused by uneven air side cooling or fan control.

The following note describes a theoretical method to assess the effect of air flow variations on ACC performance and provides simple rules to avoid potentially risky situations. An important role for safe operation plays the size of the secondary condenser. The procedure may be used as a guideline for proper sizing of the secondary condenser with respect to airflow maldistribution.

1 Introduction

A typical ACC consists of one or more parallel streets with a multitude of heat exchangers arranged in an A-frame geometry at approximately 60° base angle. The condensation is done in two steps with primary and secondary condenser (fig. 1). Heat Exchanger modules are similar in geometry. The set of heat exchangers forming a module is normally served by one fan. Steam supplied via the turbine exhaust duct enters the primary condenser section at top manifold of the exchangers and flows down the tubes to the condensate collection line ("CC line"). A considerable fraction of steam is condensed along the flow path in the primary condenser. Condensate and residual steam are collected in the CC line which conveys the remaining steam to the secondary condenser section ("dephlegmator") and condensate via the condensate line to the condensate tank. At dephlegmator inlet the remaining steam enters from below and is condensed during up-flow in the exchanger tubes. The secondary condensate is drained back to the CC line by gravity. The dephlegmator outlet connects to the inert evacuation system which keeps the vacuum permanent.

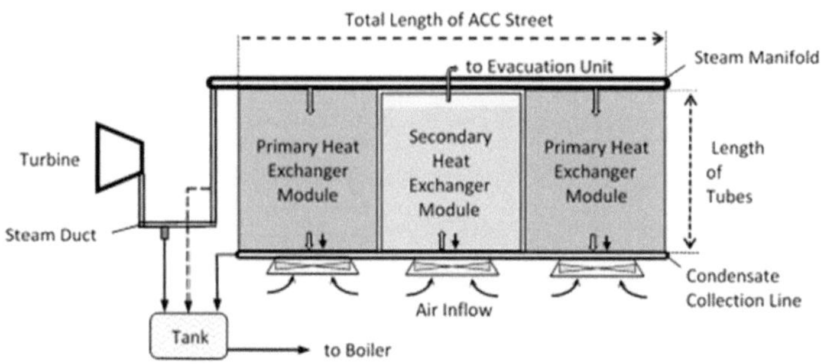

Fig. 1: General ACC Arrangement

Load adapation of the ACC is normally done by fan speed control. Variable frequency drives are typical for small ACC's. However, in large ACC units, it is sufficient to use one or two-speed fan motors. In this case, fan speed adjustment leads to stepwise cooling airflow duties resulting, in turn to discrepancies of individual module condensation capacity and different steam exit quality in primary condenser modules. Wind may have a similar effect if ACC modules are not evenly affected. Depending on arrangement individual modules may encounter poor fresh air supply. Therefore, the ACC always faces some degree of airside maldistribution - at least on a low scale.

To avoid negative consequences it must be ensured that all primary modules have at least some amount of remaining vapor at tube outlet because otherwise the inert gas carried along with the steam flow will be accumulated at the locus of minimum pressure (i.e. tube end). Locally, this results in a sharp temperature decline on the steam side and, over time, a so-called "cold spot" may develop. As a result, the capacity of the condensing system is reduced. At ambient frost conditions even freezing may occur which endangers the structural stability of the exchangers. Furthermore, cold spots promote tube side corrosion. The thermohydraulic details have been extensively discussed in the literature. All this applies to multi-row ACC's as well as single row type.

Air side maldistribution effects can only be counteracted by an appropriately sized dephlegmator which evacuates the primary section under all circumstances. However, as the cooling effectiveness of the secondary section is worse than that of the primary sections ACC suppliers tend to design the primary section as large as possible. It is therefore extremely important to understand if the dephlegmator has been properly sized. Dephlegmator sizing recommendations have been published covering the so-called "tube row effect" [1, 2]. The following report outlines a simple theoretical procedure to assess the potential of cold spot development caused by airside maldistribution.

2 ACC Model

For parameter definitions and abbreviations see the attachment.

Fig. 1 shows the general situation with a side view of an ACC street consisting of a multitude of adjacent primary and secondary modules. Steam enters the primary module tubes from the top and is condensed by cross-flowing air. Depending on local cooling airflow the steam exit quality from the primary section varies over the street length. All residual steam is conveyed by the condensate collection line ("CC line") to the dephlegmator(s) where the terminal condensation takes place in condensation reflux mode and inert gases are extracted from the top end. The condensate is accumulated in the CC line and connected to the condensate tank for permanent drain.

The current investigation is focused on the primary condenser section. Generally, the distribution of cooling airflow over the street is arbitrary leading to different condensation rates of parallel tubes or exchangers. The larger the cooling airflow duty the higher the condensation rate will be which results in enlarged steam flow into the tubes. As a first simplification we assume that cooling capacity dominates the local steam flow so that steam flow momentum changes in manifold and CC line have practically no effect. This assumption allows a simplified ACC layout where all primary modules are arranged in a consecutive line (fig. 2) – leaving out internal dephlegmator sections.

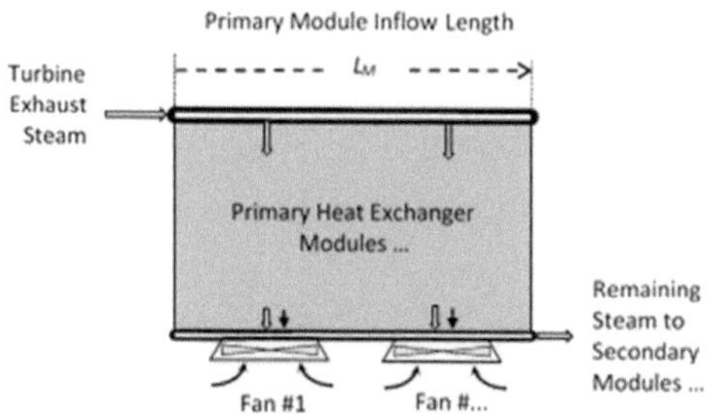

Fig. 2: Re-arranged ACC Primary Modules

With neglect of momentum effects in manifold and CC line the pressure difference top to bottom remains the same over the whole street length. If all tubes are cooled at the same equal rate the condensation capacity is constant. With varying cooling air speed each tube performs individually as local condensation duty depends on local cooling airflow. It must be made sure that in no condition the condensation end is reached within any primary module tube so that a cold spot evolves.

Given the simplifications above the air velocity profile may be re-arranged without affecting the total condensation duty. The profile starts at the lowest air speed and rises gradually along coordinate 'z' to the maximum (fig. 3). This modification allows to identify a location $z=\beta$ which divides the range of thermally fully active tubes from partly thermally active ones – the latter tubes characterized by no exit steam flow and cold spot development.

Steam enters each tube with initial speed $w(z)$ and leaves it with terminal speed $u(z)$. At cold spot formation the terminal condensation point is located within the tube at distance $l(z)$. Local cooling air velocity $c(z)$ is taken as an average over the tube length. The thermally passive region is located between end of active tube length and tube exit.

3 Boundary Conditions

The base ACC calculation model has been described elsewhere (e.g. §5 of the VGB acceptance test standard [3]). We assume that an ACC has been designed for ideal conditions – i.e. an even airflow profile. In the following the ambient air temperature does not change so that airside process and transport properties will not change. To restrict the solution to the maldistribution effect the total airside flowrate will always be kept constant with all flow profiles. These air flow profiles will be considered as imposed boundary condition.

The overall heat transfer coefficient is dominated by cooling air velocity [1, 2]. All adjacent tubes are modeled as a combined straight channel of constant width along coordinate 'z'. Momentum effects in manifold and CC line will be neglected so that initial temperature difference ("*ITD*"), manifold and CC line pressure as well as total steam side pressure drop are constant along coordinate 'z'.

Generally, we consider only the dry fraction of the approaching steam. Total manifold flow as well as residual steam flow to the dephlegmator and condensation heat are taken constant. Thus, the heat duty of the primary condenser does not change. Cooling capacity is deteriorated as soon as the air velocity profile becomes uneven. This would lead to local increase of residual steam flow which the secondary condenser cannot handle. Therefore, to support condensation capacity the manifold pressure must rise until the nominal condensation duty – which we consider constant for the purpose of this investigation – is reached. Variation of steam pressure and *ITD* will therefore be variable in the general procedure. Note that this investigation is restricted to the performance of primary condensers alone – possible variations of secondary condensation capacity will be excluded for sake of simplicity.

4 Saturation Line

Manifold pressure and condensation temperature defining *ITD* are related via the steam saturation line. Assuming small relative variations of steam pressure this relation may be linearized to

$$\left(\frac{t}{t_0}\right) \cong 1 + \frac{\vartheta_0}{t_0} \cdot S \cdot (P - 1) \tag{1}$$

with $$S \equiv \frac{3888{,}11}{\vartheta_0 \cdot [23{,}3084 - \ln(p_{M0})]^2}$$

In a similar way *ITD* ratio and manifold pressure are related as

$$P \cong 1 + S^{-1} \cdot \left[\left(\frac{\vartheta}{\vartheta_0}\right) - 1\right] \tag{2}$$

5 Heat Balance of Primary Section

Heat is exchanged along the active tube length. The small loss of saturation temperature caused by pressure changes within the tube will be similar in all adjacent tubes. Therefore, if the air flow profile is constant the exchanged heat duty anywhere in the thermally active region 'I' is the same. Assuming all tubes fully active at design conditions (i.e. no passive region) the condensed steam duty may be expressed by the difference of inlet and outlet steam velocity.

Per unit length of the primary section the heat duty is

$$\frac{\dot{q}_0}{L_M} = [\rho_{M0} \cdot w_0 - \rho_{C0} \cdot u_0] \cdot \frac{A}{L_M} \cdot r$$

The relative heat balance of the primary section may be written as

$$\frac{\dot{q}(z)}{\dot{q}_0} = \frac{\rho_M \cdot w(z) - \rho_C \cdot u(z)}{\rho_{M0} \cdot w_0 - \rho_{C0} \cdot u_0} \cdot \frac{r}{r_0} = \frac{c(z) \cdot l(z)}{c_0 \cdot L_T} \cdot \frac{\Phi(z)}{\Phi_0} \cdot \left(\frac{\vartheta}{\vartheta_0}\right)$$

With the definition of a combined airside boundary function

$$B(z) \equiv C(z) \cdot \frac{\Phi(z)}{\Phi_0}$$

we find $$W(z) - U(z) = \Lambda(z) \cdot B(z) \cdot \left(\frac{\vartheta}{\vartheta_0}\right) \tag{3}$$

In this equation the groups are defined as

$$U(z) \equiv \frac{\rho_C \cdot u(z)}{\rho_{M0} \cdot w_0 - \rho_{C0} \cdot u_0}$$

$$W(z) \equiv \frac{\rho_M \cdot w(z)}{\rho_{M0} \cdot w_0 - \rho_{C0} \cdot u_0}$$

For relative exchanger effectiveness the simplification of [3] may be used with

$$\Phi(z) = 1 - \mathrm{e}^{-NTU(z)}$$

$$NTU(z) \cong NTU_0 \cdot C(z)^{m-1}$$

6 Steam Side Pressure Drop

For pressure drop we use the simplified correlation of [1] – however, modified for mass fluxes instead of velocities. Assuming pre-dominant manifold steam pressure the pressure drop is

$$\Delta p = p_M - p_C \cong \frac{\xi \cdot l(z)}{12 \cdot d_h} \cdot v_M \cdot [\rho_M \cdot w(z) + \rho_C \cdot u(z)]^2$$

where $$\frac{v_M}{v_{M0}} \cong \left(\frac{t}{t_0}\right) \cdot P^{-1}$$

This leads to $$W(z) + U(z) \cong \kappa \cdot \Lambda(z)^{-0,5} \tag{4}$$

with $$\kappa = \kappa_0 \cdot \left(\frac{t}{t_0}\right)^{-0,5} \cdot \sqrt{\left(\frac{\Delta p}{\Delta p_0}\right) \cdot P}$$

and $$\kappa_0 \equiv W_0 + U_0 = \frac{1+D}{1-D}\,; \quad D \equiv \frac{\rho_{C0} \cdot u_0}{\rho_{M0} \cdot w_0}\,; \quad P \equiv \frac{p_M}{p_{M0}}$$

The pressure drop ratio at constant CC line pressure is

$$\left(\frac{\Delta p}{\Delta p_0}\right) = \frac{p_M - p_C}{p_{M0} - p_{C0}} = \sigma \cdot (P - P_C)$$

with $$\sigma \equiv \frac{p_{M0}}{\Delta p_0}; \quad P_C \equiv \frac{p_C}{p_{M0}}$$

7 Solution Regions

Equations (3) and (4) enable to calculate inlet and outlet steam mass fluxes. Two solution regions must be considered - active as well as partly active zone (cf. fig. 3).

- "I" fully active zone with $\Lambda_I(z) = 1$ and $z \leq \beta$:

$$U_I(z) = 0,5 \cdot \kappa - 0,5 \cdot B(z) \cdot \left(\frac{\vartheta}{\vartheta_0}\right) \tag{5}$$

$$W_I(z) = 0,5 \cdot \kappa + 0,5 \cdot B(z) \cdot \left(\frac{\vartheta}{\vartheta_0}\right) = U_I(z) + B(z) \cdot \left(\frac{\vartheta}{\vartheta_0}\right) \tag{6}$$

- "II" partially passive zone at $\beta < z < 1$ and $\Lambda_{II}(z) < 1$ where $U_{II}(z) = 0$:

$$W_{II}(z) = \Lambda_{II}(z) \cdot B(z) \cdot \left(\frac{\vartheta}{\vartheta_0}\right) = \kappa^{2/3} \cdot B(z)^{1/3} \cdot \left(\frac{\vartheta}{\vartheta_0}\right)^{1/3} \tag{7}$$

$$\Lambda_{II}(z) = \left[\frac{\kappa}{B(z)}\right]^{2/3} \cdot \left(\frac{\vartheta}{\vartheta_0}\right)^{-2/3} \tag{8}$$

At the intersection point where $U_I(\beta) = 0$ the following equation holds:

$$\kappa = B(\beta) \cdot \left(\frac{\vartheta}{\vartheta_0}\right) \tag{9}$$

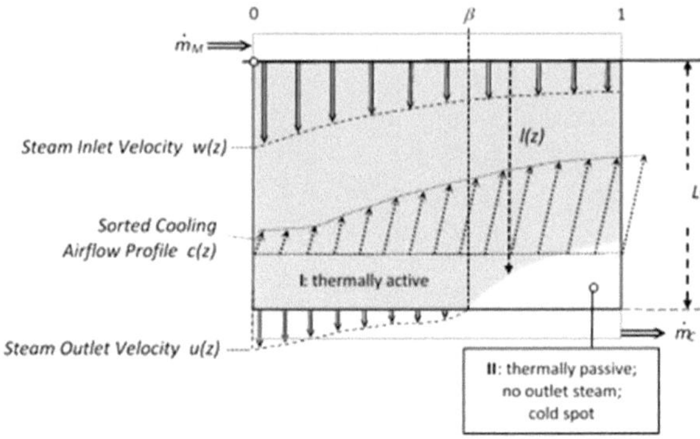

Fig. 3: Sorted ACC Primary Module System

8 Steam Flow

All steam exiting region "I" must be condensed in the dephlegmator, region "II" does not contribute.

$$\int_0^\beta U_I(z)\,dz = U_0 = 0{,}5 \cdot \kappa\,\beta - 0{,}5 \cdot \left(\frac{\vartheta}{\vartheta_0}\right) \cdot \int_0^\beta B(z)\,dz$$

∴ $$2\,U_0 = \kappa\,\beta - \left(\frac{\vartheta}{\vartheta_0}\right) \cdot J_I(\beta) \tag{10}$$

with $J_I(\beta) \equiv \int_0^\beta B(z)\,dz$

Similarly, the total turbine exhaust steam flow is

$$\int_0^\beta W_I(z)\,dz + \int_\beta^1 W_{II}(z)\,dz = W_0$$

∴ $$\left(\frac{\vartheta}{\vartheta_0}\right) \cdot J_I(\beta) + \kappa^{2/3} \cdot \left(\frac{\vartheta}{\vartheta_0}\right)^{1/3} \cdot J_{II}(\beta) = 1 \tag{11}$$

with $J_{II}(\beta) \equiv \int_\beta^1 [B(z)]^{1/3}\,dz$

Inserting equation (9) into (10) and (11) we get two equations for β and *ITD*. Solving for *ITD*:

$$\left(\frac{\vartheta}{\vartheta_0}\right) = 2\,U_0 \cdot \{B(\beta) \cdot \beta - J_I(\beta)\}^{-1} \tag{12}$$

$$\left(\frac{\vartheta}{\vartheta_0}\right) = \left\{J_I(\beta) + [B(\beta)]^{2/3} \cdot J_{II}(\beta)\right\}^{-1}$$

After re-arrangement the implicit equation for β is

$$\beta = (2U_0 + 1) \cdot \frac{J_I(\beta)}{B(\beta)} + 2U_0 \cdot J_{II}(\beta) \cdot B(\beta)^{-1/3} \tag{13}$$

The location of the intersection point depends only on the air velocity profile and basic design *NTU*. Meaningful results will only show if $\beta \leq 1$ – i.e. a cold spot in the primary condenser starts to develop. This point may be used as minimum dephlegmator fraction limit. Using the simplified exchanger effectiveness [3, 4] the condition for onset of a cold spot will be

$$D \leq \frac{C(1)\cdot(1-\Gamma)+\Gamma\cdot C(1)^m - J_I(1)}{C(1)\cdot(1-\Gamma)+\Gamma\cdot C(1)^m + J_I(1)} \tag{14}$$

In this equation we have used the approximation of [1] by replacing

$$\Phi(z)/\Phi_0 \cong 1 - \Gamma + \Gamma \cdot C(z)^{m-1}$$

where $\Gamma = NTU_0/[e^{NTU_0} - 1]$.

For evaluation the effective air velocity profile must be known.

9 Cold Spot

The active length of region "II" is found by equation (8) using interface point (9). With that the relative size of a cold spot may be calculated from

$$A_{CS} = 1 - \beta - \int_\beta^1 \Lambda_{II}(z)\, dz = 1 - \beta - B(\beta)^{2/3} \cdot \int_\beta^1 B(z)^{-2/3}\, dz \tag{15}$$

10 Steam Side

With the intersection point identified (13) all missing variables may be found:

- *ITD* – equation (12)
- Manifold pressure – equation (2)
- Saturation temperature – equation (1)
- Intersection point group κ – equation (9)
- Pressure drop $\left(\frac{\Delta p}{\Delta p_0}\right) = \left(\frac{\kappa}{\kappa_0}\right)^2 \cdot \left(\frac{t}{t_0}\right) \cdot P^{-1}$
- CC line pressure $P_C = P - \sigma^{-1} \cdot \left(\frac{\Delta p}{\Delta p_0}\right)$

11 General Airside Velocity Profile

At this point the effective air profile must be introduced. Whenever an uneven air side velocity profile develops the overall thermal efficiency deteriorates causing the manifold pressure to rise. Consequently, there is an un-wanted loss of vacuum even in the case of constant total airflow. To identify the maldistribution effect only the velocity profiles must fulfil the following condition:

$$\int_0^1 C(z)\, dz = 1$$

The following velocity profile conforms with the restrictions of the simplified ACC model:

$$C(z) = a + b\, z^R$$

with $$a = \frac{1+R}{n+R} > 0; \;\; b = \frac{(1+R)\cdot(n-1)}{n+R} > 0; \;\; R > 0$$

$$n = \frac{c_2}{c_1} > 1$$

This general form allows profiles with narrow low speed as well as narrow peak speed range. Fig. 4 shows three air velocity profile examples with different peak speed ratio and coordinate exponent to generate the desired velocity trend.

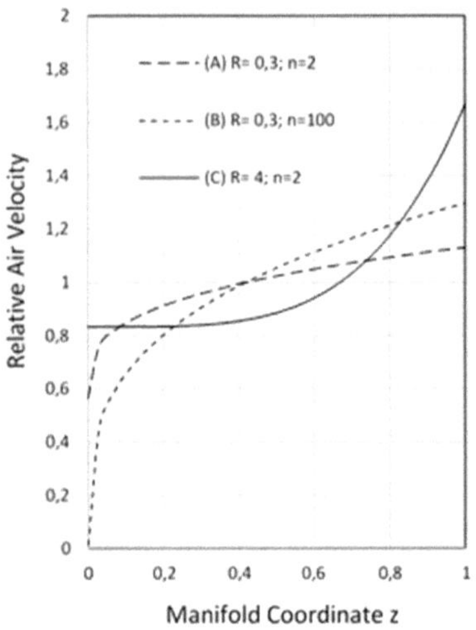

Fig. 4: Standard Air Flow Profiles

Modules in small ACC's consisting of only one street are similarly affected by external wind, i.e. the air velocity profile is close to horizontal on a large ACC fraction. There may be only a small section with noticeable loss of cooling - variant (A). In medium sized ACC's this section may be larger especially with increased wind speed – variant (B). At high wind speed local loss of cooling may be encountered which puts the peak speed ratio to an extreme. Variant (C) is an example where a small ACC proportion encounters very good cooling with the rest of modules in similar cooling condition. This may apply to large multi-street ACC's with a marked proportion of circumferential modules encountering poorer supply of air compared to internal modules. Therefore, we shall distinguish between:

(A) Small ACC, moderate wind: $R = 0{,}3;\ n = 2$

(B) Extreme wind, locally blocked cooling: $R = 0{,}3\ ;n = 100$

(C) Large ACC, moderate speed profile: $R = 4\ ;\ n = 2$

Note that hot air recirculation caused by wind is cannot be modeled by this procedure because the restrictions used for the proposed procedure do not apply.

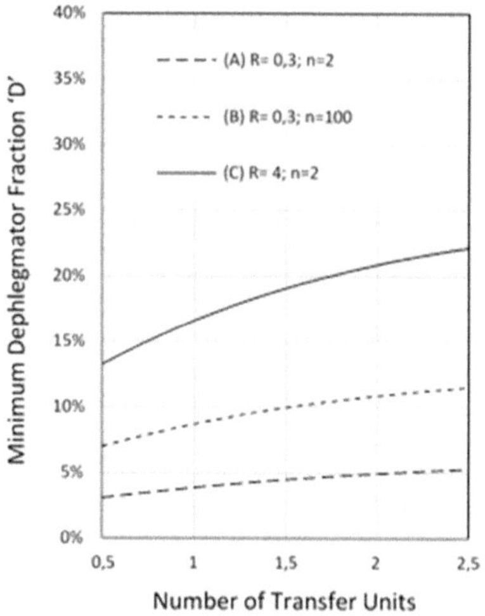

Fig. 5: Minimum Dephlegmator Fraction

The minimum dephlegmator fraction *'D'* (14) is shown in fig. 5. Note that contemporary large ACC's are characterized by *'D'* ratios down to 1 : 6 or 1 : 7. It can be seen that apart from the air flow profile *'D'* depends only on design *NTU*. As for *NTU* – severe sound restrictions compel low air velocities which, in turn lead to large values, typically up to $NTU \approx 1{,}8$.

Maldistribution of type (A) or (B) with small low speed sections pose no risk in operation because minimum *'D'* remains small. This is different in variant (C) with no low speed section. Even a moderate speed ratio may require an effective *'D'* of at least 1 : 5 or less.

Note that in multi-row condensers the so-called "tube row effect" may – depending on tube type sequence - impose *'D'* fractions larger than that of airside maldistribution. Interestingly, the main parameter for the "tube row effect" is again *NTU* [1]. On any account the best strategy is to operate the dephlegmator always at peak load to suppress cold spot formation in the primary condenser. This is put into practice by appropriate fan speed control.

To assess the potential size of a cold spot further input parameters are needed (15). Fig. 6 shows an example for *NTU* = 1,8 (low noise design), 10 kPa manifold pressure, 1,5 kPa primary condenser pressure drop and 20 K *ITD*. The examples show that the affected area provoked by airside maldistribution is small.

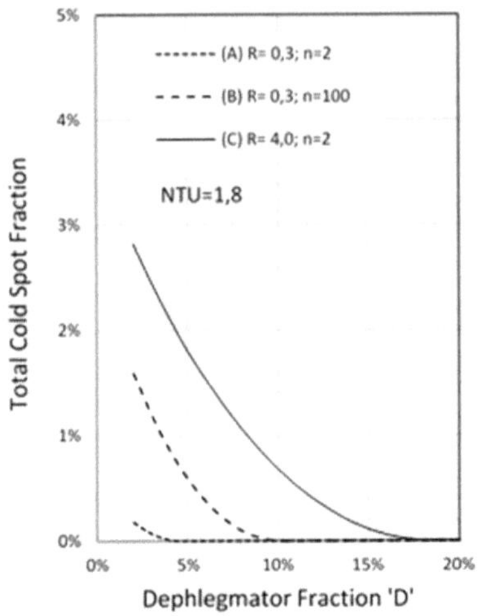

Fig. 6: Cold Spot Fraction

(NTU = 1,8; 10 kPa; Δp = 1,5 kPa; ITD = 20 K)

Cold spot formation goes along with increase of manifold pressure which reduces the plant efficiency. In practice, increased steam pressure cannot be used as sole indicator of airside maldistribution because other reasons such as surface degradation, hot air recirculation or vacuum leaks may have a larger effect.

12 Fan Speed Control Primary Condenser

The outlined procedure can also be used to model different fan speed settings of parallel primary condenser modules. This is typical for control of ACC cooling capacity. Simplifying, we assume a step-wise air velocity profile comprising two ranges with (a) low fan speed and (b) high fan speed (fig. 6). Air velocity is taken approximately proportional to fan speed. As before, airside maldistribution will be compared to constant airflow (e.g. variable frequency control).

The form of the two-step profile conforming with our model is

$$C_1(z) = \frac{1}{\beta_a + n\cdot(1-\beta_a)} = C_1 \qquad 0 \le z \le \beta_a$$

$$C_2(z) = \frac{n}{\beta_a + n\cdot(1-\beta_a)} = C_2 \qquad \beta_a < z \le 1$$

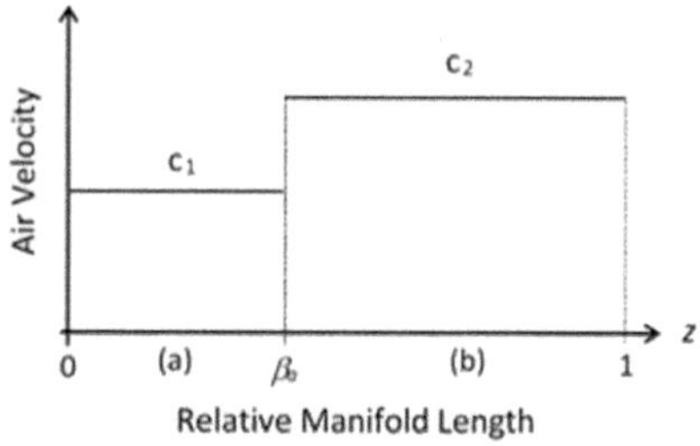

Fig. 7: Two-Step Fan Control

If an intersection point exists it must be at the end point of the low speed fan range $\beta = \beta_a$ (fig. 7). In this case, equation (9) holds with

$$\kappa = C_2 \cdot \frac{\Phi(C_2)}{\Phi_0} \cdot \left(\frac{\vartheta}{\vartheta_0}\right) = B_2 \cdot \left(\frac{\vartheta}{\vartheta_0}\right)$$

By re-arranging equations (10) and (11) we find

$$\left(\frac{\vartheta}{\vartheta_0}\right) = \frac{2U_0}{\beta_a\cdot(B_2 - B_1)} = \frac{W_0}{B_2 - 0{,}5\,\beta_a\cdot(B_2 - B_1)}$$

so that the minimum dephlegmator fraction may be expressed as

$$D \ge \frac{0{,}5\,\beta_a\cdot(B_2 - B_1)}{B_2 - 0{,}5\,\beta_a\cdot(B_2 - B_1)} \qquad (16)$$

Again, we find a dependency on *NTU* which is hidden in the *'B'* functions. In linearized form these functions are

$$B_1 \cong (1-\Gamma)\cdot C_1 + \Gamma \cdot {C_1}^m$$

$$B_2 \cong (1-\Gamma)\cdot C_2 + \Gamma \cdot {C_2}^m$$

Results of equation (16) are shown on figures 8 to 10. As could be expected the smaller the speed step the better is the situation. For typical ACC designs with design *'D'* ratios of 1 : 6 or 1 : 7 cold spots will not develop at fan speed ratio of 80% or 67%.

Generally, the larger the *NTU* or, the lower the fan speed ratio the worse the situation. Low noise designs with fan speed ratio of 50% (fig. 10) or less show a serious tendency to develop cold spots especially if the low speed fraction is large (80% or higher).

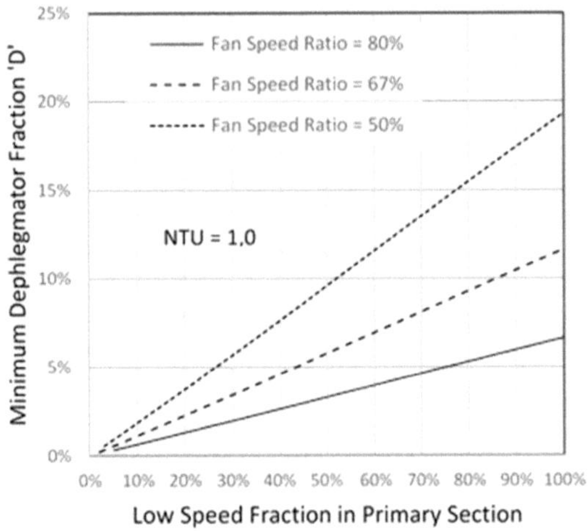

Fig. 8: Minimum Dephlegmator Size, Two-Step Fan Switch, NTU low

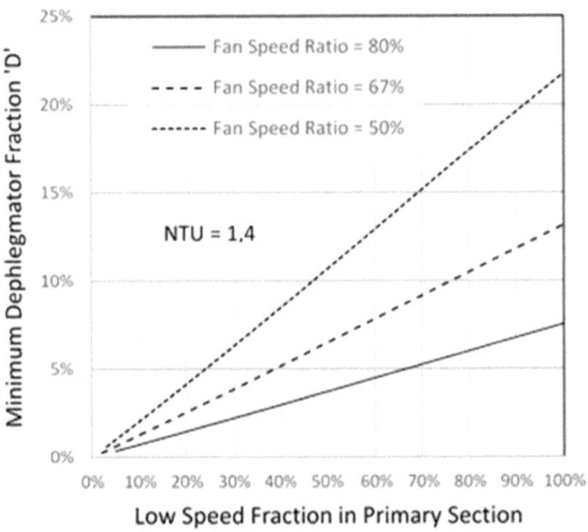

Fig. 9: Minimum Dephlegmator Size at Two-Step Fan Switch, NTU normal

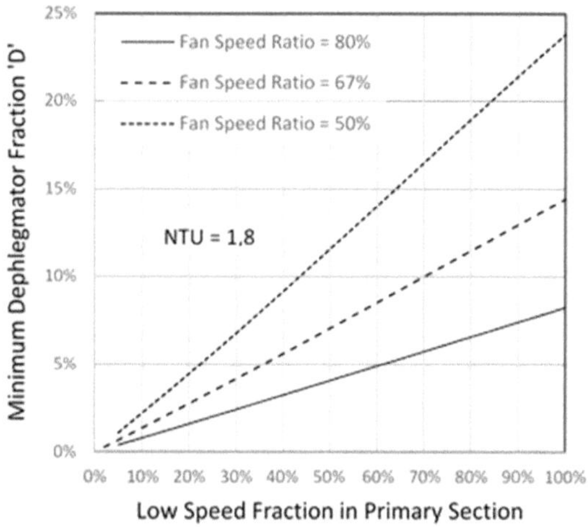

Fig. 10: Minimum Dephlegmator Size, Two-Step Fan Switch, high NTU

If the design *'D'* ratio is in the range of 1 : 4 or 1 : 5 (e.g. multi-row ACC's with non-Aluminum finning) development of cold spots is highly improbable. If this goes along with low design *NTU* (fig. 8, 9) definitely no cold spot will evolve as a result of airside maldistribution.

13 Fan Speed Effect on Steam Pressure

To find the effect of fan speed control on steam pressure equation (9) does not hold any more. Basically, we have to differentiate between the existence of a cold spot and a fully active primary condenser. To decide if a cold spot exists equation (14) may be used.

If no cold spot evolves equations (5) and (6) apply to the whole length co-ordinate. Assuming constant total steam flow at inlet and outlet we find

$$\left(\frac{\vartheta}{\vartheta_0}\right)_{active} = 1/[B_2 - \beta_a \cdot (B_2 - B_1)] \tag{17}$$

If, on the other hand, a cold spot develops equations (5) to (7) must be considered. At steam outlet

we have $\quad U_0 = 0{,}5\,\beta_a \cdot [\kappa_0 \cdot X - B_1] \cdot \left(\frac{\vartheta}{\vartheta_0}\right)$

and at inlet $\quad W_0 = 0{,}5\,\beta_a \cdot [\kappa_0 \cdot X + B_1] \cdot \left(\frac{\vartheta}{\vartheta_0}\right) + \kappa_0^{2/3} \cdot J_{II}(\beta_a) \cdot \left(\frac{\vartheta}{\vartheta_0}\right) \cdot X^{2/3}$

with $\quad X = \left(\frac{\vartheta}{\vartheta_0}\right)^{-0{,}5} \cdot \sqrt{\left(\frac{\Delta p}{\Delta p_0}\right) \cdot \frac{P}{t/t_0}} \quad (18)$

$$J_{II}(\beta_a) = [B_2]^{1/3} \cdot (1 - \beta_a)$$

The auxiliary group has been defined to enable linearization as $X^{2/3} \cong \frac{1}{3} + \frac{2}{3} \cdot X$. After re-arrangement we end up with

$$X = \frac{3\beta_a \cdot B_1 + 2\, J_{II}(\beta_a) \cdot U_0\, \kappa_0^{-1/3}}{3\, \beta_a - 4\, J_{II}(\beta_a) \cdot U_0\, \kappa_0^{-1/3}}$$

and
$$\left(\frac{\vartheta}{\vartheta_0}\right)_{passive} = \frac{2\, U_0}{\beta_a \cdot (\kappa_0 \cdot X - B_1)} \qquad (19)$$

Depending on the applicable situation equations (17) or (19) enable to assess the effective manifold steam pressure via equation (2). Figures 11 to 13 show the increase of steam pressure for a typical design configuration (dephlegmator fraction 1 : 7, manifold steam pressure 10 kPa, *ITD* 20 K). Generally, the pressure effect remains negligible if the low speed fraction is small or extremely high.

If no cold spot develops the largest pressure effect is in the 50% - 60% range of low speed fans with the result nearly independent of *NTU*. Although there is no critical accumulation of inert gas - power plant output is negatively affected. Especially, at low fan speed ratio (50%) un-wanted loss of vacuum increases distinctly. From that point of view fan speed switches should be moderate to avoid loss of efficiency.

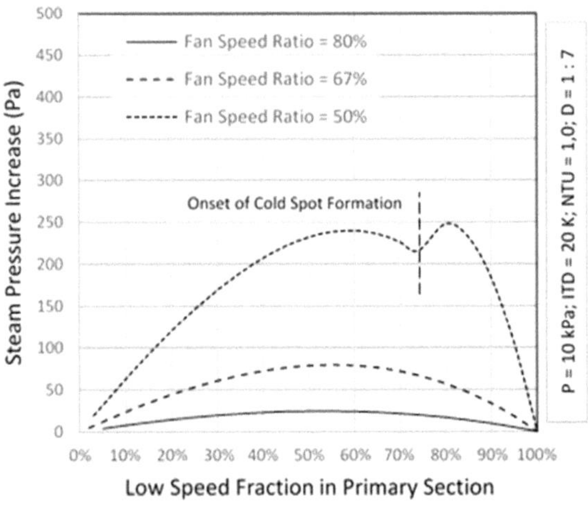

Fig. 11: Effective Pressure Increase – low NTU

As soon as a cold spot develops (cf. fig. 11 to 13) the steam pressure increases quickly - the peak level now dominated by *NTU*. The cold spot peak in the selected example is nearly doubled by changing *NTU* from 1 (fig. 11) to 1.8 (fig. 13). This is an extra minus point on top of the risk of frost formation.

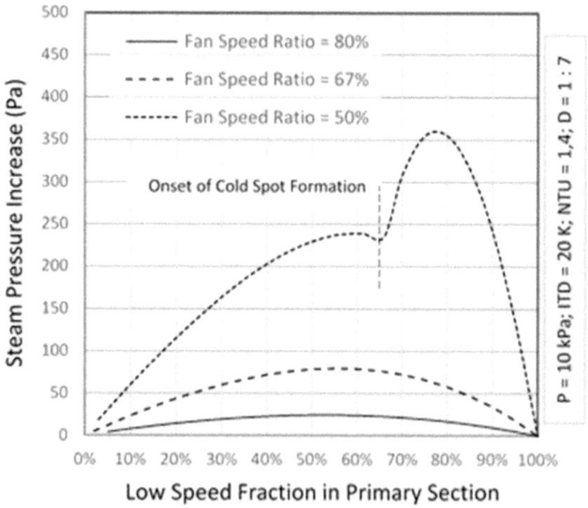

Fig. 12: Effective Pressure Increase – normal NTU

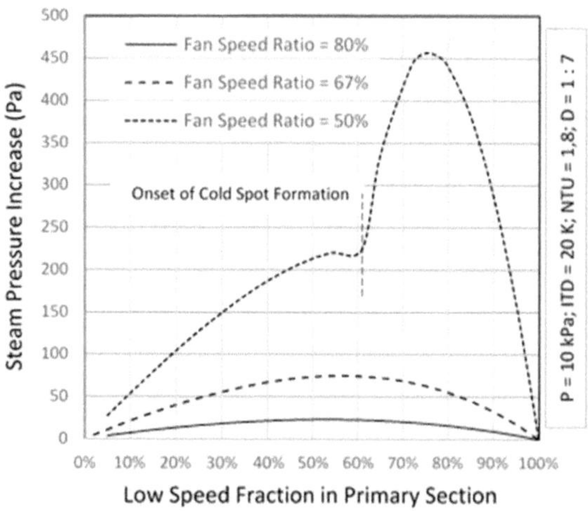

Fig. 13: Effective Pressure Increase – high NTU

As to dephlegmator fraction it should be noted that the effective heat duty proportion *'D'* is generally smaller than the total exchanger surface proportion. This means practically, that *'D'* fraction 1 : 7 corresponds to a dephlegmator module ratio of 1 : 6. So the example above and the conclusions drawn cover a multitude of concurrent ACC designs.

14 Cold Spot Counteraction

What can be done in case of cold spot development ? A good strategy takes into account that cold spots can only be formed after some time of operation. In practice – with good vacuum tightness – it takes hours to fill the cold spot volume with inert gas. If full speed and part speed fans are inter-changed from time to time the cold volume will be forced out of the affected section and the process starts from the beginning but, in different modules of the ACC. With this kind of fan switching sequence fatal results of ACC operation can be avoided. Still, loss of power plant efficiency remains. There is no way of trying to counteract airside maldistribution already on the design stage.

An effective measure is to foresee frequency controlled fan motors which allow concurrent switching of primary module fans so that all fans are running at same speed. However, natural airside velocity variations cannot be counteracted by this provision. To ensure nominal ACC performance under all circumstances there is no way but to over-design the ACC to cover at least some proportion of the estimated loss of vacuum. Alternatively, installation of local air guide vanes may be envisaged based on extensive CFD analysis. Another option to safeguard even cooling flowrate – especially fit for large ACC arrays – is to equip the externally arranged modules with larger fan motors to make up for the extra loss caused by peak air approach velocities.

15 Conclusion

The present investigation is focused on the effect of un-even air velocity profiles compared to theoretically constant air speed. To this end, a simple theoretical model dealing with the mal-distribution effect on the performance of ACC's has been developed. Either natural velocity variations (ambient or arrangement effects) may give rise to cold spot formation in the primary modules or, different fan speed settings.

In any case, if cold spots are formed or not - loss of power plant efficiency will be the result. Recommendations for sizing the secondary condenser to avoid cold spots caused by airside maldistribution are made using prototype profiles covering a wide range of possible scenarios. Although these profiles have been arbitrarily selected general conclusions could be drawn with confidence.

The procedure allows to estimate possible maldistribution effects already on the design stage. Examples of some typical ACC designs are given.

Glossary

β	-	intersection point of 'z' coordinate, start of passive zone
β_a	-	range of 'z' coordinate with low speed fan(s), step profile
Δp	-	tube side pressure drop manifold – CC line (Pa)
ξ	-	steam duct resistance coefficient
ρ_C	-	steam density (kg/m^3) tube bottom, CC line
ρ_M	-	steam density (kg/m^3) top of tubes (manifold)
ϑ	-	initial temperature difference: steam - air (K)
Γ	-	auxiliary group [3], [4]
$\kappa_0 = \frac{1+D}{1-D}$	-	auxiliary group design point (-)
κ	-	auxiliary group (-)
$\Lambda(z) = l(z)/L_T$	-	relative thermally active length (-)
$\Phi(z)$	-	exchanger effectiveness (-)
X	-	auxiliary group, equation (18)
A	-	total steam side area of primary modules (m^2)
A_{CS}	-	cold spot percentage
$a,\ b$	-	air velocity profile constants
$B(z)$	-	extended air velocity function (-)
c_0	-	design air velocity (m/s) – straight air flow profile
$c(z)$	-	local air velocity (m/s)
$C(z) = c(z)/c_0$	-	relative local air velocity (-)
d_h	-	tube side hydraulic diameter (m)
$D = \dot{m}_{C0}/\dot{m}_{M0}$	-	dephlegmator (steam flow) ratio
$J_I(\beta)$	-	integral expression, equations (10) to (14)
$J_{II}(\beta)$	-	integral expression, equation (11), (13), (19)
L_M	-	total manifold length covering primary modules (m)
L_T	-	tube length (m)
$l(z)$	-	local thermally active length (m) of primary modules
$m \cong 1/3$	-	average airside Reynolds exponent [3]
$\dot{m}$	-	dry steam flow (kg/s)
$M_M = \dot{m}_M/\dot{m}_{M0}$	-	relative dry steam inlet flowrate
$M_C = \dot{m}_C/\dot{m}_{C0}$	-	relative dry steam flowrate exiting CC line
$n = c_2/c_1$	-	ratio peak to lowest air velocity (or: inverse of fan speed)
NTU_0	-	number of transfer units (reference point)
p_C	-	absolute steam pressure (Pa) CC line level (tube outlet)

p_M	-	absolute steam pressure (Pa) manifold level (tube inlet)
P	-	relative manifold steam pressure, equation (2)
P_C	-	relative CC line steam pressure
$\dot{q}$	-	heat duty (W)
r	-	condensation heat (J/kg)
R	-	air profile exponent
S	-	auxiliary group saturation line
t	-	steam temperature (°Cel), saturation line
$u(z)$	-	outlet steam velocity at CC line (m/s)
$U(z)$	-	relative steam outlet mass flux (CC line)
v_M	-	steam specific volume (m^3/kg)
$w(z)$	-	inlet velocity at manifold (m/s)
$W(z)$	-	relative steam inlet mass flux (manifold)
z	-	relative coordinate along manifold L_M, $0 \leq z \leq 1$

Indices

0	-	reference (average) condition, no mal-distribution
M	-	manifold (top of ACC)
C	-	condensate collection line (CC line)
I	-	range $0 \leq z \leq \beta$
II	-	range $\beta < z \leq 1$

Bibliography

[1] H.G. Schrey, J. Kern: "Zum Rohrreiheneffekt bei gasbeaufschlagten Kondensatoren"(„On the tube row effect at gas cooled condensers"), Int. J. Heat & Mass Transfer 24 (1981), 335 – 342 (in German)

[2] H.G. Schrey: „Zur Bestimmung des Leistungsverhaltens bei mehrreihigen luft- bzw. gasgekühlten Kondensatoren mit Inerten" („Determination of operating performance of multi-row air-cooled condensers with inerts"), Heat and Mass Transfer 18 (1984) No. 3, 185-191 (in German)

[3] "VGB Guideline Acceptance Test Measurements and Operation Monitoring of Air-Cooled Condensers under Vacuum", VGB R-131 Me, published by "VGB Technische Vereinigung der Grosskraftwerksbetreiber", 1997

[4] "Air-Cooled Steam Condensers, Performance Test Codes", ASME PTC 30.1 - 2007, The American Society of Mechanical Engineers, 2008